Will Black Holes Devour the Universe?

&
100 Other
Questions
& Answers
About Astronomy

by Melanie Melton
from the publishers of ASTRONOMY magazine

Clear Skies!
Melanie Melton

KALMBACH BOOKS

ACKNOWLEDGMENTS

Dave Alexander, Greg Novacek, Becky Cooper and Nancy Luis, my friends and co-workers at the Lake Afton Pubic Observatory; John Foley—the computer graphics expert; "Chairman" Corkie Martin, Sue Peterson, Nancy Vaught, Pat and John Davis and Leslie and Ed Willett. Thanks, Les! Robert and Helen Melton, my parents.

DEDICATION

In memory of my brother Craig, who took pride in asking the difficult questions.

Illustrations: Loke Yuen Fung and Eric Jegen.

Printed in the United States of America.

Library of Congress Cataloging-in-Publication Data

Melton, Melanie.
    Will black holes devour the universe? : and 100 other questions & answers about astronomy / Melanie Melton. — 1st ed.
       p. cm.
       Includes bibliographical references.
       ISBN 0-913135-20-8

    1. Astronomy—Popular works. 2. Astronomy—Miscellanea. I. Title.

QB44.2.M45 1994   520
                                    QBI94-1410

Contents

Introduction

Questions about the Sun

1. What is the Sun?
2. How far away is the Sun?
3. Is the Sun burning?
4. How hot is the Sun?
5. What is the Sun made of?
6. If the Sun is made of gas, could you fall right through it?
7. Will our Sun ever burn out?
8. Do we really need the Sun?
9. Could we replace the Sun with something else if the Sun suddenly died?
10. Will humans replace the Sun when it dies?

Questions about the Planets

11. Where did the planets get their names?
12. Which is the biggest planet?
13. Which is the smallest planet?
14. Is there life on other planets?
15. Is there a tenth planet?
16. Does every planet have a moon?
17. Which planet has the most moons?
18. Which planet has the largest moon?
19. Why is Mars sometimes called the Red Planet?

30. What formed the asteroid belt?
31. Why do stars fall?
32. Why do meteors leave a streak of light across the sky?
33. What is the difference between a meteor and a meteorite?
34. What are meteorites made of?
35. What is a comet?
36. Do comets blaze across the sky?
37. What's the big deal about Halley's Comet?

QUESTIONS ABOUT EARTH

38. What does Earth have to do with astronomy?
39. Why does the Sun move across the sky?
40. What causes the seasons?
41. Why is it cold in winter and hot in summer?
42. If it's winter in North America, why is it summer in Australia?

QUESTIONS ABOUT THE MOON

43. What is the Moon made of?
44. How was the Moon formed?
45. How old are rocks on the Moon's surface?
46. Why does the Moon change shape?
47. What is a lunar eclipse?
48. Can you always see the Moon?
49. What is the man in the Moon?
50. Does the same side of the Moon always face Earth?
51. What is the far side of the Moon?
52. Is there a side of the Moon that is always dark?

QUESTIONS ABOUT SPACECRAFT AND SPACE TRAVEL

53. Can the space shuttle travel to the Moon?
54. Does the space shuttle have to dodge stars?
55. Why should we build a space station?
56. If we wanted to, could we travel to the Moon tomorrow?
57. How many astronauts have walked on the Moon?
58. Can we see footprints on the Moon?
59. If we build colonies on the Moon, could we see them from Earth?
60. Why haven't humans traveled to other planets?
61. Should we travel to Mars?
62. Can we travel to the stars?
63. Do any of the spacecraft we send to the planets come back to Earth?
64. How do we get all of those pretty pictures of the planets?

QUESTIONS ABOUT STARS

65. Is it true that stars twinkle and planets don't?
66. What is a constellation?
67. How many constellations are there?
68. Do astronomers study constellations?
69. Can you see a constellation through a telescope?
70. Why do the stars only come out at night?
71. How many stars can you see in the night sky?
72. What is the North Star?
73. Is there a South Star?
74. Will the North Star always show us the way north?
75. Is the North Star the brightest star in the sky?
76. What is a light-year?
77. What is the closest star to Earth?
78. How does a star form?
79. Will a star live forever?
80. How long do stars live?
81. What is a red giant?
82. What is a white dwarf?
83. What is a supernova?
84. What is a pulsar?
85. What is a neutron star?

QUESTIONS ABOUT BLACK HOLES

86. What is a black hole?
87. What is a singularity?
88. What is an event horizon?
89. How do black holes form?
90. Since black holes are black, how do you find them?
91. What would happen if our Sun became a black hole?
92. Could you travel through a black hole?
93. Can we send a spaceship through a black hole?
94. Where does everything go after it enters a black hole?
95. Will black holes devour the universe?

QUESTIONS ABOUT GALAXIES

96. What is a galaxy?
97. What is the Milky Way?
98. When can I see the Milky Way?
99. Do we have a picture of our Galaxy?
100. Are all galaxies shaped like a pinwheel?
101. How many galaxies are there?

INTRODUCTION

Will black holes devour the universe? Is it true that stars twinkle and planets don't? Is the North Star the brightest star in the sky? Will our Sun ever burn out?

Welcome to the world of astronomy, where questions like these are asked all the time! Astronomy is a fascinating field of study filled with stars many times larger than our Sun, planets larger than Earth and smaller than our Moon, exploding stars, stellar nurseries, and much more. Astronomy is an integral part of our lives as well. The Sun in our sky is a star and Earth, our home, is a planet. It is natural to want to know more about the world we live in. It is also natural to question how we fit in with everything else in the universe.

Astronomy tries to answer these questions. With the answers comes the realization that the universe we live in is a strange and exotic place, where facts are sometimes stranger than fiction.

Speaking of answers, you may be curious about the questions at the beginning of the page. The answers are: no, no and yes, no, and yes (but don't lose any sleep over it). We'll explore these questions and more as we begin our journey into the world of astronomy. Enjoy!

IMPORTANT DEFINITIONS

Before we delve into the many common questions asked in astronomy, it's a good idea to start with a few basic definitions. Some of these may seem a bit obvious, but it's better to review common knowledge than to assume everyone knows what I'm talking about.

A STAR:
A huge ball of hydrogen and helium gas that gives off tremendous amounts of energy by nuclear fusion.

Notice that the definition doesn't say anything like "can only be seen at night." That is because the star closest to us can only be seen during the day. More about that later.

A PLANET:

A large, spherical object that orbits a star.

Planets are usually the largest objects traveling around a star, following a predictable path. This predictable path is called an orbit. Compared with planets, other objects that travel around a star—comets and asteroids, for example—are small and irregularly shaped.

Although planets are large, they are still much smaller than stars. You could stack more than 100 Earths across the diameter of the Sun, our own star.

Besides size, another difference between planets and stars is energy. A star gives off its own energy, while a planet or moon merely reflects the energy of a star.

Did you know that the only reason we can see Jupiter, Venus, and the other planets is because of the Sun? Sunlight strikes these planets and bounces back toward Earth. If the Sun didn't shine on the other planets, we wouldn't see them.

10 20 30 40 50 60 70 80 90 100

100 EARTHS

1 SUN

A MOON:

A naturally occurring object that orbits a planet. "Naturally occurring" is an important part of this description. Earth has one naturally occurring Moon. It travels around Earth once every 29 days.

However, Earth has thousands of artificial moons. Since the Soviet Union launched Sputnik, the first satellite, in 1957, Earthlings have been creating their own moons. Every weather, communication, spy, or other satellite launched into orbit around Earth could be considered a moon—an object in orbit around our planet. In this book I will use the word moon to mean natural satellite.

THE SOLAR SYSTEM:

Everything within the Sun's gravitational influence.

Our solar system includes the Sun, nine planets (including Earth), comets, meteors, asteroids, and perhaps much more. The Latin name for our Sun is Sol. Solar system, or the Sun's system, is the name given to those things that are affected by the Sun.

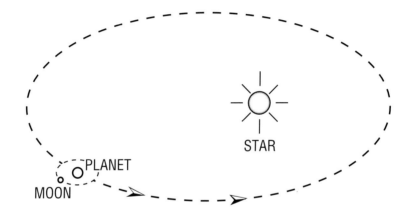

A BRIEF SUMMARY

A star is a huge sphere of gas that gives off its own energy. A planet orbits a star and a moon orbits a planet. All of these are parts of a solar system. Now that you are going around in circles (sorry, I couldn't resist) maybe we should settle down and get to the questions.

QUESTIONS

ABOUT

THE SUN

1. WHAT IS THE SUN?

THE ANSWER:

The Sun is a star.

SIMPLE EXPLANATION:

Because the Sun looks so different from the stars you see at night, many people don't realize that our Sun is a star. The difference between the nighttime stars and the Sun is not size or brightness. Our Sun is average; many stars are larger and many are smaller. The main difference between our Sun and the other stars is distance.

SOME MORE DETAILS:

Imagine you are outside at night with two of your friends. Each friend has the same type of flashlight. One stands 3 feet from you, and the other stands 20 feet away. Both shine their flashlights in your eyes. (Some friends!) Which of the two flashlights would appear brighter? The one closest to you, of course. Is it really brighter? No, it is just closer. This is why the Sun appears so much brighter in our sky than the other stars. It's a lot closer.

2. HOW FAR AWAY IS THE SUN?

THE ANSWER:

The Sun is, on average, 93,000,000 miles (150,000,000 kilometers) from Earth.

SIMPLE EXPLANATION:

93 million miles (150 million kilometers) is a long distance away. From Earth, it would take a spacecraft traveling 23,000 miles per hour

(36,800 kilometers per hour)—a realistic cruising speed—almost six months to reach the Sun. In comparison, the next nearest star system, called Alpha Centauri, is almost 25,000,000,000,000 miles (40 trillion kilometers) away. It would take the same spacecraft over one hundred thousand years to reach Alpha Centauri.

So the Sun may seem as though it is really far away, but it is in our backyard compared with even the next-closest star.

3. IS THE SUN BURNING?

THE ANSWER:

Not like a fire in the fireplace. Comparing the way the Sun is burning to a fire in the fireplace is like comparing a fireplace to a hydrogen bomb. The first will warm a few people with its heat, the other will level a large city. They are two totally different types of reactions.

SIMPLE EXPLANATION:

A fire in the fireplace uses wood for fuel. The burning of the wood is a chemical reaction.

The Sun's fuel is hydrogen atoms. The Sun produces energy by fusing together hydrogen atoms to form helium. This fusion is a nuclear reaction, and nuclear reactions are much more powerful than chemical reactions. Millions of such nuclear reactions are going on all the time in the Sun's core.

SOME MORE DETAILS:

Nuclear fusion requires very special conditions. In fact, the conditions are so special that the only place in the solar system where nuclear reactions occur naturally is the center of the Sun.

First of all, fusion requires hydrogen. Since the Sun is made of almost 80 percent hydrogen, that requirement is the easiest to meet.

A second requirement is a temperature of several million degrees Fahrenheit. Finally, you need extremely high pressures, millions of times greater than we are used to on Earth.

When hydrogen fuses into helium the process releases a tremendous amount of energy. The energy gradually makes its way to the Sun's surface and out into space. Earth receives this energy mostly in the form of heat and light.

4. HOW HOT IS THE SUN?

THE ANSWER:

It depends on which part of the Sun you are talking about. Temperatures range from thousands to millions of degrees F.

SIMPLE EXPLANATION:

The surface of the Sun is about 10,000°F (5,500° Celsius). Sunspots, dark areas on the Sun's surface, are cooler—only 8,000°F (4000°C). As you go deep into the Sun, the temperature increases. At the core, the temperature rises to several million degrees F.

5. WHAT IS THE SUN MADE OF?

THE ANSWER:

Hydrogen, helium, and a few trace elements.

SIMPLE EXPLANATION:

75 percent of the Sun's matter is hydrogen and 24 percent is helium.

The remaining 1 per cent consists of oxygen, carbon, nitrogen, silicon, magnesium, neon, and other elements.

6. IF THE SUN IS MADE OF GAS, COULD YOU FALL RIGHT THROUGH IT?

THE ANSWER:

No. There is nothing solid to stand on, even if you could survive the extremely hot temperatures. But that doesn't mean that you would fall right through it, either.

SIMPLE EXPLANATION:

If you could magically survive the Sun's heat, the pressure you would encounter as you sank deeper into the gas would crush you long before you made it to the center of the Sun.

SOME MORE DETAILS:

Have you ever dived to the bottom of the deep end of a swimming pool and felt your eardrums hurt? The deeper you go under the surface, the greater the pressure. Even under 10 feet of water, you notice a definite increase in pressure. The water is squeezing you from all sides.

Deep sea explorers require tanks called submersibles made of extremely thick metal to survive the tremendous pressures found a couple of miles beneath the sea. Without the protective gear, they would be crushed. The Sun is 880,000 miles in diameter. Imagine trying to go 440,000 miles below its surface! If humans have trouble surviving the pressures found a few miles beneath Earth's oceans, how could anyone possibly survive 440,000 miles beneath the surface of our Sun? The pressure would be so tremendous that you would be squashed as you got started. So even though the Sun is made of gas, there is no way anyone, or anything, could pass through it.

7. WILL OUR SUN EVER BURN OUT?

THE ANSWER:

Yes. Don't lose any sleep over it, though. Astronomers figure the Sun will last another 5 billion years.

SIMPLE EXPLANATION:

If you know your car uses a gallon of gasoline for every 20 miles you drive, and you know your gas tank holds 20 gallons, you can predict that you will go 400 miles before you run out of gas.

Astronomers use the same reasoning to determine how long our Sun will continue to give off energy. They know nuclear reactions are taking place in the Sun, and they know how massive the Sun is. In other words, they know how much fuel the Sun has and how fast it is using that fuel. So they can predict how long the Sun will fuse hydrogen before it "runs out of gas." It has been "on" for almost 4.6 billion years and has at least that long to go before it dies.

8. DO WE REALLY NEED THE SUN?

THE ANSWER:

Absolutely, positively YES! Everything, from the food we eat to the house we live in, is affected by the Sun.

SIMPLE EXPLANATION:

The energy to grow all the food we eat comes from the Sun. Plants need the Sun to grow. As they grow, plants convert carbon dioxide in the atmosphere into oxygen that we breathe. Some of these plants are eaten by animals and humans. Animals like cows, sheep, and chickens are then eaten by humans.

The food chain is not the only thing affected by the Sun. The house you live in and the books you read depend on the Sun too. The wood in your house came from trees, which need sunlight to grow. Trees also provide us with paper, boxes, and other supplies. And they provide wood for fires to warm us on chilly nights.

But wood fires would not provide enough heat to support life without the Sun. Without sunlight, no matter how big a fire you built, the Earth would be a very cold place at -300°F (-184°C).

Earth would also be very dark. We would see no Moon in the sky, since moonlight is actually reflected sunlight. Only the pinpoints of distant starlight would break up the blackness. But we wouldn't see them, either, because without sunlight, there would be no life on Earth to witness the darkness.

9. WILL THE SUN JUST BURN OUT AND GO BLACK?

THE ANSWER:

No. The Sun will go through a series of contractions and expansions before it is no more.

SIMPLE EXPLANATION:

When the Sun has used up most of its fuel, it will expand to become a red giant (a huge, cool star). Our red giant Sun will be so big that it will incinerate Mercury and Venus and nearly reach Earth. The Sun will be cooler, but since the surface of the Sun will be much closer to Earth, Earth will get really hot. All of the water in the oceans will boil away and life on Earth will be exterminated.

After it uses up all of its fuel, the Sun will start to collapse. As it collapses, it will shed its outer layer, forming what is called a planetary nebula. As the outer layer moves away from the Sun, the Sun will continue to shrink until it is about the size of Earth. Then it will be what is called a white dwarf. From there, the white dwarf Sun will gradually fade away.

Our Sun, now an average, yellow star, will become a red giant, then a planetary nebula, and then a white dwarf before it fades to black.

10. WILL HUMANS REPLACE THE SUN WHEN IT DIES?

THE ANSWER:

No. When the Sun dies, there won't be any humans around to worry about it, at least not on Earth.

SIMPLE EXPLANATION:

As the Sun uses up the last of its fuel, it will swell in size. As it expands, the distance between Sun and Earth will shrink and the temperature on our planet will soar. All life will die and the oceans will boil away. Don't worry, though—we have almost four billion years of smooth sailing before things begin to heat up.

QUESTIONS

ABOUT THE

PLANETS

11. WHERE DID THE PLANETS GET THEIR NAMES?

THE ANSWER:

From the ancient Greeks and Romans.

SIMPLE EXPLANATION:

The word "planet" is derived from the Greek word meaning "wanderer." The ancients noticed that most of the stars in the sky stayed fixed in their constellations, but a few of those "stars" moved, or wandered, across the sky. Because they did not understand why these "stars" were moving across the sky, they decided these heavenly bodies must have something to do with the gods. So they named the planets after the gods and goddesses of the time. The ancient Greeks named the planets. The Romans later adopted the Greek mythology, and today we know the planets by their Roman names.

SOME MORE DETAILS:

Of all the planets, Mercury moves the fastest through the sky. Because of this, Mercury was named after the fleet-footed messenger god. (Mercury is the closest planet to the Sun and orbits once every 88 days.)

Venus was named after the goddess of beauty because its brightness dominates the early morning or evening sky.

Mars was named after the god of war because of its blood-red appearance.

Jupiter was named after the king of the gods. Its brightness does not equal that of Venus, but it seems to dominate the heavens as its orbit takes it slowly across the sky.

Saturn, the slowest-moving of the visible planets, was named after the king of the Titans (a mythical race of giants).

Ancient people named only Mercury, Venus, Mars, Jupiter and Saturn. The more distant planets—Uranus, Neptune, and Pluto—

cannot be seen without a telescope and were not discovered until much later. When naming the more distant planets, astronomers continued the ancient tradition. Uranus was a son of Mother Earth and one of the first gods. Neptune was god of the seas, and Pluto was the ruler of the Underworld.

12. WHICH IS THE BIGGEST PLANET?

THE ANSWER:
Jupiter.

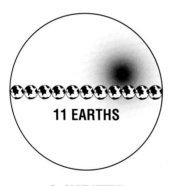

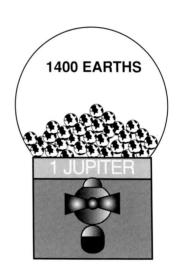

SIMPLE EXPLANATION:
Jupiter is by far the largest planet in the solar system. Seventy per cent of all of the mass in the solar system—outside of the Sun, that is—is found within Jupiter.

Jupiter is over 11 times larger than Earth. That means you could fit 11 Earths across Jupiter's diameter. If Jupiter were hollow, you could fit 1,400 Earths inside it.

13. WHICH IS THE SMALLEST PLANET?

THE ANSWER:
Pluto.

SIMPLE EXPLANATION:
The most distant planet is also the smallest. Pluto is even smaller than Earth's Moon. Pluto is 1,380 miles (2,300 kilometers) in diameter, while our Moon is 2,086 miles (3,476 kilometers) across. Because of its distance from the Sun—almost 4 billion miles (6.6 billion kilometers)—and its small size, Pluto was not discovered until 1930. The ninth planet cannot be seen with the naked eye. Even through a telescope, Pluto appears only as a faint spot.

14. IS THERE LIFE ON OTHER PLANETS?

THE ANSWER:
Not that we know of.

SIMPLE EXPLANATION:
For life as we know it to exist, certain requirements must be met. Some of these are: temperature that is not too hot or too cold, the presence of liquid water, and just the right amount of oxygen.

Spacecraft have flown by or landed on every planet in the solar system except Pluto. They have found no traces of life. Two spacecraft, Viking 1 and 2, took soil samples on the Martian surface and tested them for life-forms. The results were negative.

SOME MORE DETAILS:

While every planet in the solar system has its own beauty, as far as we know Earth is the only one suited for life. Mercury has no atmosphere, and the temperature varies from 660°F (350°C) during the day to -274°F (-170°C) at night. Quite a change!

On Venus, you would have to survive passage through sulfuric acid clouds, an average temperature of 900°F (470°C) and an atmospheric pressure 90 times greater than Earth's. On Mars, the temperature isn't quite as harsh, with highs around 70°F (20°C) and lows around -220°F (-140°C). However, the Martian atmosphere is not the least bit friendly to humans. Its air is made almost entirely of carbon dioxide and its atmospheric pressure is less than 1/100 of Earth's.

The gas giant planets—Jupiter, Saturn, Uranus, and Neptune—are all extremely cold. They are made up mostly of hydrogen and helium gases and have no solid surface to stand on. Living on these planets would be difficult indeed.

Many kinds of life might survive under much harsher conditions than we have on Earth, but for now the only life we can find in the solar system is right here.

15. IS THERE A TENTH PLANET?

THE ANSWER:

Astronomers have not found one yet.

SIMPLE EXPLANATION:

A planet beyond Pluto would be extremely difficult to find because of its vast distance from the Sun. Pluto is 4 billion miles (6.6 billion kilometers) from the Sun and is only 1,380 miles (2,300 kilometers) across. A tenth planet would be farther away from the Sun than Pluto, so it would be much fainter and even more difficult to see.

SOME MORE DETAILS:

Astronomers have been looking for a tenth planet because they were puzzled by the orbits of Uranus, Neptune, and Pluto. These three planets didn't follow their predicted paths. Recently, astronomers have recalculated their orbits. Using the more accurate measurements of mass and size of Uranus and Neptune collected from the Voyager 2 spacecraft, the orbital discrepancies disappear. It seems that planets were right where they were supposed to be, and it was the calculations that were incorrect.

16. DOES EVERY PLANET HAVE A MOON?

THE ANSWER:

No.

SIMPLE EXPLANATION:

In our Solar System the two innermost planets, Mercury and Venus, have no moons.

SOME MORE DETAILS:

As you move farther from the Sun, the planets have more and more moons. Earth has 1 moon, Mars has 2, Jupiter has 16, and Saturn has 19. Then the numbers start to drop off. Uranus has 15 moons, Neptune has 8, and Pluto has 1.

Some books you read may give different numbers of moons for the gas giant planets. As Voyager 1 and 2 flew by Jupiter (1979) and Saturn (1981) and Voyager 2 continued on to Uranus (1986) and Neptune (1989), many new moons were discovered.

Some books claim that Saturn has as many as 23 moons. That may be possible, but before a moon can be officially added to the list of satellites, there must be at least two photographs of it to prove that it is a moon with a predictable orbit and not just a part of Saturn's rings. It is easy to prove that the larger objects are moons, but many of the recently discovered moons are very small.

The numbers of moons listed above are current as of 1994.

Those numbers may change as more spacecraft visit the planets or as old spacecraft images are further studied.

17. WHICH PLANET HAS THE MOST MOONS?

THE ANSWER:
Saturn.

SIMPLE EXPLANATION:
Saturn has 19 moons that have been officially identified. There may be more moons hidden within Saturn's rings.

SOME MORE DETAILS:
Saturn's largest moon, Titan, is 3,090 miles (5,150 kilometers) in diameter. Titan is the second-largest moon in the solar system and larger than the planets Pluto and Mercury. For comparison, our Moon measures 2,086 miles (3,476 kilometers) across.

Mimas, Enceladus, Tethys, Dione, Rhea, Hyperion, Iapetus, and Phoebe are the names of some of Saturn's other moons.

18. WHICH PLANET HAS THE LARGEST MOON?

THE ANSWER:
Jupiter.

SIMPLE EXPLANATION:
Ganymede, the largest moon of Jupiter, is also the largest moon in the solar system. It is 3,166 miles (5,276 kilometers) in diameter and is larger than the planets Mercury, at 2,930 miles (4,878 kilometers), and Pluto, at 1,380 miles (2,300 kilometers). For comparison, our Moon is 2,086 miles (3,476 kilometers) in diameter.

SOME MORE DETAILS:
Ganymede is made of approximately half rock and half ice. Its surface is mottled with dark and light areas, and several craters can be seen. The craters appear as white marks on its surface. When meteors slammed into Ganymede to form the craters, they broke through a thin dusty layer covering the moon and revealed the white water-ice below.

Other moons of Jupiter are: Io, Europa, Callisto, Metis, Adrastea, Amalthea, Thebe, Leda, Himalia, Lysithea, Elara, Ananke, Carme, Pasiphae and Sinope. All of Jupiter's moons are named after mythical lovers or companions of Zeus. (In Greek mythology, Zeus was the king of the gods. Jupiter was the Roman name for the king of the gods.)

19. WHY IS MARS SOMETIMES CALLED THE RED PLANET?

THE ANSWER:
The surface of Mars is yellow-orange. This gives Mars a ruddy glow.

SIMPLE EXPLANATION:

The surface of Mars contains a small amount of iron. Since there is a small amount of oxygen in the Martian atmosphere, the iron begins to oxidize, or more simply, rust. The surface of Mars is behaving just like an old car. It's rusting.

20. ARE THERE CANALS ON MARS?

THE ANSWER:

No.

SIMPLE EXPLANATION:

In 1877 an Italian astronomer, Giovanni Schiaparelli, was observing Mars. He reported that he saw canali on the red planet. In Italian, canali means channels or grooves. Unfortunately, canali was translated incorrectly into English as canals.

The word "canals" brings to mind man-made trenches. It led to speculation that there were intelligent creatures living on Mars, channeling their limited water supply through these canals. An American astronomer, Percival Lowell, claimed he saw as many as 160 canals crisscrossing the Martian surface. However, few astronomers believed there were canals on Mars.

When the first spacecraft flew by the planet in the 1960s, no canals could be seen. More detailed photographs have found no sign of these mysterious canals.

SOME MORE DETAILS:

The canals may have been optical illusions caused by Earth's atmosphere. When observing Mars through a telescope, a viewer can see dark areas on some parts of the surface. These darker areas are regions of low reflectivity, not canals.

21. WHAT IS THE GREAT RED SPOT ON JUPITER?

THE ANSWER:
A huge storm similar to Earth's hurricanes.

SIMPLE EXPLANATION:
The Great Red Spot is a large storm. It varies in size from one to three times the size of Earth and in color from deep red to pale brown.

SOME MORE DETAILS:
Astronomers are not sure what caused the Great Red Spot to form or how it varies in size and color. One cause could be Jupiter's fast rotation. Earth takes 24 hours to rotate once on its axis. Jupiter is more than eleven times larger than Earth, so you would think it would rotate much more slowly, right? Wrong! Jupiter rotates once every 9 hours and 50 minutes, more than twice as fast as Earth.

The hydrogen and helium gas that make up Jupiter tend to get stirred up as the planet spins around once every ten hours. Although this contributes to the huge hurricane, it does not explain the storm's direct cause.

The Great Red Spot has been "storming" on Jupiter for at least 350 years. No one knows how long it was there before we discovered it or how long it will continue. If you compare it to storms on Earth, it could be around for a long time.

Hurricanes on Earth form over large bodies of warm water. They gain strength as long as they are over water. When they pass over land, their source of energy is gone and they die off. Jupiter doesn't have any "land," as we do on Earth. It will take some other force to slow down the Great Red Spot. We will have to wait and watch.

22. COULD YOU LAND A SPACECRAFT ON JUPITER?

THE ANSWER:
No.

SIMPLE EXPLANATION:
Jupiter has no "surface" to land on. A spacecraft would sink through thicker and thicker clouds until the clouds were thicker than split-pea soup. By then the pressure would be so great that it would crush the spacecraft.

SOME MORE DETAILS:
See the explanation under Question 6: "If the Sun is made of gas, could you fall right through it?" Even though the Sun is much bigger than Jupiter, the idea is the same. The deeper you go into Jupiter, the greater the pressure.

23. WHAT ARE THE RINGS OF SATURN MADE OF?

THE ANSWER:
Ice and rocks.

SIMPLE EXPLANATION:

Scientists are not sure how the rings of Saturn formed or even how they remain in orbit around the planet. Before Voyager 1 and 2 flew by Saturn, scientists thought the rings were continuous bands. Instead, the spacecraft detected thousands of rings filled with icy rocks anywhere from the size of a grain of sand to the size of a house. Flying through the rings of Saturn would be like flying through a blizzard.

SOME MORE DETAILS:

One of the reasons the rings of Saturn appear so bright in a telescope is that they consist of chunks of water ice and ice-covered rocks. Have you ever gone outside on a sunny day after a snowstorm? The sunlight reflecting off the snow is so bright that it can give you a bad headache. Sunlight reflecting off the ice in the rings of Saturn may be just as bright, but we are too far away for it to give us a headache.

24. IS SATURN THE ONLY PLANET THAT HAS RINGS?

THE ANSWER:

No.

SIMPLE EXPLANATION:

All four of the gas giant planets (Jupiter, Saturn, Uranus, and Neptune) have rings, but Saturn's are by far the biggest and brightest.

SOME MORE DETAILS:

Jupiter has one lone ring that consists of very dark material. The ring can't be seen from Earth. Astronomers didn't even know if Jupiter had any rings when the first Voyager spacecraft flew by the planet. They took a chance and aimed the camera of the spacecraft where they thought any rings might be. As luck would have it, the camera did capture a picture of one lonely, dark ring. Astronomers knew that Uranus had rings before Voyager 2 flew by the planet. Not because they could see them, but because they couldn't see through them. Sounds strange, doesn't it?

Astronomers were observing Uranus as it passed in front of a distant star. (This is called an occultation.) They expected to see the star disappear when Uranus passed in front of it. What they didn't expect to see was the star blinking on and off before Uranus got to it. After Uranus moved away from the star, the star blinked on and off again. Something surrounding Uranus was blocking the light from that star. This something turned out to be nine dark, narrow rings.

Astronomers observed Neptune during an occultation to see if they could discover rings around that planet. What they discovered was confusing. Sometimes a star would blink on and off, and sometimes it wouldn't. They decided that Neptune must have ring segments instead of complete rings.

When Voyager 2 sent back pictures, it showed that Neptune did indeed have complete rings—three of them, in fact. The rings were just too thin to block the light from a star. Scientists also found that the outer ring contained uneven clumps. The clumps were thick enough to block the starlight during an occultation, and that was what astronomers on Earth had observed.

QUESTIONS ABOUT ASTEROIDS, METEORS, COMETS, AND OTHER SPACE JUNK

25. WHAT IS AN ASTEROID?

THE ANSWER:

A rock made of stone, iron, or a combination of both that travels around the Sun.

SIMPLE EXPLANATION:

Asteroids come in many different shapes and sizes. Mostly they look like what they are—chunks of rock and metal. They are not round like planets, although they are sometimes called "minor planets" because they travel around the Sun. In many cases, their orbits are predictable. For example, the asteroid Vesta travels around the Sun once every 3.63 years. In comparison, Earth travels around the Sun once every 365.25 days, or one year.

Asteroids are called "minor" because they are much smaller than the planets. Asteroids range in size from less than half a mile (1 kilometer) to over 600 miles (1,000 kilometers) in diameter. In comparison, Pluto, the smallest planet, is 1,350 miles (2,300 kilometers) in diameter.

26. WHAT IS THE ASTEROID BELT?

THE ANSWER:

An area between the orbits of Mars and Jupiter where thousands of asteroids are found.

SIMPLE EXPLANATION:

Astronomers have officially named and numbered over 3,000 asteroids within the main asteroid belt, although over 5,000 have been discovered in photographs. This number may seem large, but astronomers believe there are even more out there. They estimate there are roughly 100,000 asteroids between the orbits of Mars and Jupiter.

Even though there are so many of them, asteroids are not easily seen because they are so small. Of the 5,000 asteroids discovered, only about 230 are larger than 60 miles (100 kilometers) in diameter. Most of them are less than half a mile (1 kilometer) across.

27. WHAT IS THE LARGEST ASTEROID?

THE ANSWER:

Ceres.

SIMPLE EXPLANATION:

Ceres was the first asteroid to be discovered, probably because it is the largest of all the known asteroids. It is almost 600 miles (1,000 kilometers) in diameter and takes 4.6 years to travel around the Sun.

SOME MORE DETAILS:

Ceres is definitely in a minority when it comes to size. Only two other asteroids, Pallas and Juno, come close to its size. Both are slightly larger than 180 miles (300 kilometers) in diameter. Astronomers have charted 230 other asteroids larger than 60 miles (100 kilometers) across. The vast majority of the other 5,000 main-belt asteroids are less than half a mile (1 kilometer) across.

28. Is it dangerous to travel through the asteroid belt?

THE ANSWER:

Not really.

SIMPLE EXPLANATION:

When scientists sent the first spacecraft to visit Jupiter, they were concerned about the asteroid belt. Could a spacecraft travel through this crowded portion of space and survive, or would it be pulverized by an asteroid? Unfortunately, there was no way to avoid the belt.

To the surprise of some of the scientists, nothing happened.

As the spacecraft entered the belt, none were pulverized. In fact, Voyager 1 and 2, which had instruments on board to record impacts on the spacecraft, detected no increase in the number of impacts as they traveled through the belt. It was as if the asteroid belt didn't exist. Is this strange? Not really.

SOME MORE DETAILS:

Even though there may be over 100,000 asteroids in the asteroid belt, there is still quite a bit of empty space within the belt.

This band, approximately 140 million miles (223 million kilometers) wide between Mars and Jupiter, covers an enormous area. The asteroids orbit within an area of over 265,000,000,000,000,000 square miles (442,000,000,000,000,000 square kilometers). Each asteroid has an average area of over 2.6 trillion square miles (4.4 trillion square kilometers all to itself. With an average of more than 1.5 million miles (2.5 million kilometers) between asteroids, they are relatively easy to dodge. So the asteroid belt is not really a dangerous part of space to fly through, after all.

29. ARE ALL ASTEROIDS FOUND WITHIN THE ASTEROID BELT?

THE ANSWER:

No. The asteroids that are found within the asteroid belt are called "main-belt" asteroids. At least three other types are found elsewhere in our solar system.

SIMPLE EXPLANATION:

The Trojan asteroids share the same orbit as the planet Jupiter. They travel in front of Jupiter or trail after the giant planet. Other asteroids have very elliptical, or egg-shaped, orbits that bring them into the inner solar system. The Amor asteroids cross the orbit of Mars, and the Apollo asteroids cross the orbit of Earth.

30. WHAT FORMED THE ASTEROID BELT?

THE ANSWER:

Astronomers aren't really sure, but they have a good idea. They believe asteroids are the material left over from the formation of the solar system.

SIMPLE EXPLANATION:

If Jupiter hadn't been so close and large, this leftover material might have formed a planet. But Jupiter's gravity is so strong that it kept tugging on all of the small pieces, never allowing them to group together.

SOME MORE DETAILS:

When the asteroid belt was first discovered, many astronomers thought that the asteroids might have come from a planet that had broken up or shattered. However, there are some problems with this idea.

One problem is the amount of material within the asteroid belt. If you put all of the asteroids together, the resulting planet would be much smaller than Pluto. This "planet" doesn't fit in with the rest of the solar system.

Another solution astronomers can't accept is the idea that a planet exploded. Today most astronomers have given up on the idea of a shattered planet and agree that asteroids are leftover debris from the formation of the solar system.

31. WHAT IS THE DIFFERENCE BETWEEN A FALLING STAR, A SHOOTING STAR, AND A METEOR?

THE ANSWER:

They are all different names for the same thing: a piece of rock or dust that burns up as it enters our atmosphere.

SIMPLE EXPLANATION:

A so-called "falling star" or "shooting star" has nothing to do with the stars you see at night. The nighttime stars are light-years distant. They are huge balls of hydrogen and helium gas, just like our Sun.

A falling or shooting star is only a fragment of rock that runs into Earth's atmosphere. (Sorry, I know the definition isn't very romantic, but that's the way it is.) That tiny piece of rock is called a meteor.

A meteor starts out as a piece of rock or dust floating in outer space. As Earth travels around the Sun, it sometimes runs into these tiny rocks and dust particles. When they fall into Earth's atmosphere, they burn up, converting some of their mass into heat and leaving a bright streak across the sky.

32. WHY DO METEORS LEAVE A STREAK OF LIGHT ACROSS THE SKY?

THE ANSWER:

Meteors burn up as they pass through Earth's atmosphere, putting on a brilliant display. This display is caused by something that everyone is familiar with, although you may not realize it. That something is friction.

SIMPLE EXPLANATION:

What do you do when your hands are cold? If there is not a fire or heater to warm them by, you usually rub them together. Try it now—rub them together for a few seconds. They're warm, aren't they?

It was the friction caused by rubbing them together that warmed them up. The same friction, on a larger scale, causes that bright streak of light in the night sky.

Even if you use hand lotion to keep your hands smooth, each is covered with patterns of ridges and grooves. These ridges can resemble a miniature mountain range. When you rub your hands together, these ridges bump into each other and resist the movement. You easily overcome their resistance, but the result is heat. The faster and harder you rub your hands together, the more resistance you encounter. The greater the resistance, the more friction, and the more friction, the more heat. The friction between two sticks of wood will start a fire if you rub them against one another fast enough.

A meteor is traveling so fast when it rubs against the Earth's atmosphere that both the atmosphere and the meteor act as rough surfaces against one another. The trail of light you see in the night sky is the hot air left behind as the meteor burns its way through the atmosphere.

SOME MORE DETAILS:

Sometimes meteors are large enough that they don't burn up completely as they pass through the atmosphere. The surviving rock strikes the ground and forms a crater.

33. WHAT IS THE DIFFERENCE BETWEEN A METEOR AND A METEORITE?

THE ANSWER:

One is up in the air and one is on the ground. The difference all has to do with location.

SIMPLE EXPLANATION:

A meteor is a rock traveling through Earth's atmosphere. If the same rock makes it all the way down to the ground, it becomes a meteorite. So if you see one in the sky, it's a meteor. Once it hits the ground, it becomes a meteorite.

34. WHAT ARE METEORITES MADE OF?

THE ANSWER:

Stone and iron.

SIMPLE EXPLANATION:

There are three major types of meteorite: stony, stony-iron, and iron. The most common type of meteorite is stony. But the type people find most often is iron. Does that make sense? Yes. Read on and you will see why.

SOME MORE DETAILS:

If you see a meteor in the sky and watch it hit the ground, chances are it will be a stony meteorite. Stony-type meteors make up about 93 percent of all the meteors. Yet they are very difficult to find unless you actually see one strike the ground nearby, and this is very unlikely.

Stony meteorites are difficult to find because of their appearance. They look like Earth rocks. If you were plowing a field and stumbled across a stony meteorite, you would probably just push it aside with a few grumbles and not notice that it was a space rock.

If, however, you stumbled across an iron meteorite (which make up only 6 percent of the meteor population) you would know something was strange. These meteors are made up of 80–90 percent iron and 10–20 percent nickel. In other words, they are heavy! If you tried to kick one of these "rocks" out of the way, you would end up stubbing your toe and not budging the "rock." You are much more likely to recognize the iron meteorite than the stony one.

The third type of meteorite is stony-iron, which is made up partly of stone and partly of iron. This type of meteorite is the rarest, making up only 1 percent of all meteors.

35. WHAT IS A COMET?

THE ANSWER:
A dirty snowball that orbits the Sun.

SIMPLE EXPLANATION:
A comet is a large, dirty snowball that spends most of its time in the far reaches of our solar system. Millions of miles away and very dark, a comet is not much to look at. However, when its orbit brings it close to the Sun, things begin to heat up—literally. The once-dark, dirty snowball begins to vaporize.

As a comet vaporizes, a cloud of gas and dust forms around the solid snowball. A shower of particles streaming from the Sun pushes some of this gas and dust away from the comet's nucleus, forming a long, graceful tail.

SOME MORE DETAILS:
You would think the tail of a comet would always be trailing it, the way a dog's tail always trails behind the dog. That seems to make sense, but it's not the case. Sometimes the tail is behind the comet, sometimes it is in front. The tail acts as a streamer, showing the direction of the particles coming from the Sun, which astronomers call the solar wind. Since the solar wind is constantly blowing away from the Sun, the comet's tail will always point away from the Sun. When the comet is approaching the Sun, therefore, its tail is behind it. When the comet is retreating from the Sun, its tail is in front of it.

36. DO COMETS BLAZE ACROSS THE SKY?

THE ANSWER: No.

SIMPLE EXPLANATION:

Over the course of a long evening, you may see a comet move very slightly against the background of stars, but a comet will not blaze across the sky in a matter of moments.

SOME MORE DETAILS:

Some comets do travel fast in their orbits, and from night to night you may see a notable change in their position. But keep in mind that these objects are very far away from us. The farther away an object is, the slower it appears to be moving. To help you picture this, think about a jet flying overhead. That jet may be traveling at a speed of 500 miles per hour, yet from where you stand on the ground, it may look as though it is barely moving.

37. WHAT'S THE BIG DEAL ABOUT HALLEY'S COMET?

THE ANSWER:

Halley's Comet helped astronomers understand what comets are.

SIMPLE EXPLANATION:

After the appearance of an extremely bright comet in 1682, a scientist named Edmond Halley decided he would try to calculate the comet's path. In doing his research, he noted that there were several bright comets throughout history.

Some of these comets had even passed through the same region of the sky as the one in 1682. After many years of calculating, Halley finally determined that many of these different sightings, which took place every 75–76 years, were not different comets, but the same comet returning over and over again. When the comet appeared again in

1758, it proved that Halley was right. The returning comet was named in his honor.

SOME MORE DETAILS:

In the past, Halley's comet has been a spectacular sight in the night sky. However, the last time it approached the Sun, in 1986, Earth was not in a good position to observe the comet. With all of the publicity from the news media, many people were excited about its return but were disappointed when they actually saw the faint, fuzzy blob. The next time around, in the year 2062, the comet will be closer to Earth and promises to be a beautiful sight.

QUESTIONS

ABOUT

EARTH

38. WHAT DOES EARTH HAVE TO DO WITH ASTRONOMY?

THE ANSWER:

Astronomy is the study of space and everything found within it—planets, moons, stars, nebulae, black holes, comets, asteroids, and much more. The planet Earth with its Moon is one of nine planets revolving around a star.

Our star and nearby stars seen in the night sky lie within one spiral arm of our home galaxy. Two small companion galaxies lie near our own. These three galaxies share a region of space with seventeen other galaxies, forming what is called the local group. Beyond our local group of galaxies lie thousands of superclusters of galaxies.

Earth, a tiny planet orbiting an average star, is a small but integral part of the vast world of astronomy.

SIMPLE EXPLANATION:

By studying other planets, scientists can learn more about Earth than would otherwise be possible. Since our very existence depends on Earth's stable climate, good soil, and breathable air, scientists are constantly trying to learn more about our home planet. By comparing ancient volcanoes on Mars with the volcanoes in Hawaii, astronomers learn more about both. By observing impact craters and debris patterns on the Moon, Mercury, and Mars, astronomers can better identify impact regions on Earth. By studying the thick, soupy atmosphere of Venus, astronomers can gain an insight into the atmosphere of primeval Earth and the dangers facing Earth's atmosphere today, such as the ozone hole and the greenhouse effect.

SOME MORE DETAILS:

To learn more about Earth, astronomers also study the Sun. Small changes in the Sun's surface temperature could cause the next ice age or worldwide drought. Increased activity on the Sun's surface, such as solar flares, can disrupt radio and satellite communications all around the world. The more we know about our nearest star, the better we can understand our life on Earth.

So you see, astronomy is all around us.

39. WHY DOES THE SUN MOVE ACROSS THE SKY?

THE ANSWER:

The Sun only appears to be moving across the sky, rising in the east and setting in the west. What is really moving is Earth.

SIMPLE EXPLANATION:

Earth takes 24 hours to rotate on its axis. That is one day. This rotation is what causes the Sun to "travel" across the sky.

Think about the last time you rode in a car. As you traveled down the road, it looked as though things were moving outside your window while you were sitting perfectly still. As passengers on Earth, we are along for the same kind of ride through space. The only way we sense our motion is to see things "outside" the Earth—the Sun and stars— moving. As Earth turns, they seem to move in and out of view.

SOME MORE DETAILS:

A simple demonstration can help explain things.

Find a lamp in your room. (A ceiling light will work, but it is better to have a light at eye level.) For this demonstration, the light will be the Sun and you will be Earth.

Your nose can be the house you live in. Stand directly in front of the light, facing it. This represents noon. On Earth, the Sun would be straight overhead.

Now, Earth (that's you) rotates on its axis. Start turning to your right. What happens to the Sun? It appears to move. Stop when your left shoulder is nearest the Sun. This represents sunset. Keep turning. You can't see the Sun at all now. This is nighttime. Everything you see in the room now represents stars you could see at night. As you continue to spin on your axis, soon the Sun will show up over your right shoulder. This represents sunrise. Keep turning without stopping, from sunrise to sunset.

Even though it looks as if the light is moving, Earth (yourself) is the one doing all of the work.

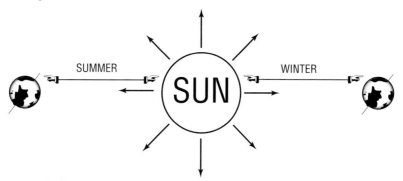

40. WHAT CAUSES THE SEASONS?

THE ANSWER:

Earth's tilted axis.

SIMPLE EXPLANATION:

During a summer in the Northern Hemisphere, it may seem that Earth is moving closer to the Sun as it gets hotter and hotter. This is not the case at all. In fact, during the summer, Earth is actually farther away from the Sun than it is during winter. Instead of distance from the Sun, the seasons are entirely the result of a tilted axis.

As Earth spins on its axis every day, it is also traveling around the Sun. (There is enough spinning going on to get everybody dizzy!)

Earth's axis is not pointed straight up and down as it travels around the Sun. It's tipped over a little.

SOME MORE DETAILS:

The tilted North Pole of Earth is pointed to a star named Polaris, or the North Star. This doesn't change as Earth moves around the Sun.

What changes is Earth's orientation to the Sun. During summer the North Pole is tilted toward the Sun. Six months later Earth has rotated, so that the North Pole is tilted away from the Sun, though still pointed at the North Star. The tilt of the axis toward or away from the Sun causes the seasons.

41. WHY IS IT COLD IN WINTER AND HOT IN SUMMER?

THE ANSWER:

Because of Earth's tilted axis.

SIMPLE EXPLANATION:

When Earth's North Pole tilts away from the Sun, the Northern Hemisphere doesn't get much sunlight, so it gets cold. When Earth's North Pole is pointed in the direction of the Sun, the Northern Hemisphere warms up and summer arrives.

SOME MORE DETAILS:

During winter in the Northern Hemisphere, Earth is actually about three million miles closer to the Sun than it is during the summer. Earth's distance from the Sun doesn't affect the seasons. It's the tilted axis that brings us snow in the winter.

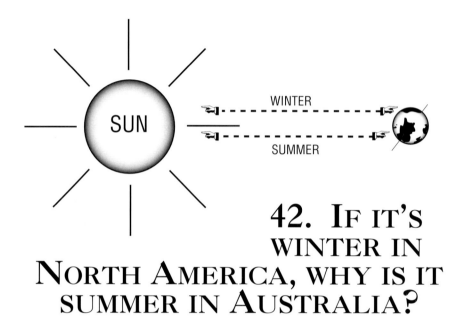

42. IF IT'S WINTER IN NORTH AMERICA, WHY IS IT SUMMER IN AUSTRALIA?

THE ANSWER: Once again, because of Earth's tilted axis.

SIMPLE EXPLANATION: During winter in North America, the North Pole is tilted away from the Sun. If the North Pole is tilted away from the Sun, where is the South Pole? Pointed toward the Sun, of course. So the seasons are reversed. It is summer in Australia.

QUESTIONS ABOUT THE MOON

43. WHAT IS THE MOON MADE OF?

THE ANSWER:
Rocks.

SIMPLE EXPLANATION:
Earth and the Moon are made of similar elements, but there are a few differences.

A big difference between Earth and the Moon is water, or lack of water. Moon rocks don't contain any water. Moon rocks also show that there was no oxygen present when they were formed. (Both water and oxygen are plentiful on Earth.) Another difference is that rocks on the Moon contain more calcium, aluminum, titanium, uranium, and thorium than Earth rocks.

SOME MORE DETAILS:
Most of the rocks on the Moon were formed by cooling lava. They are igneous and basaltic rocks. The lava did not come from volcanoes, however, because there are no volcanoes on the Moon. It came from the interior of the Moon, seeping up to the surface through cracks in the Moon's crust.

Other rocks, called breccia (breh´-cha), were formed during meteor impacts. The heat from the impacts caused rocks to melt and stick together, forming new rocks.

44. HOW WAS THE MOON FORMED?

THE ANSWER:
Scientists don't really know, since they can't travel back through

time and watch. But using clues from the past, they have pieced together what probably happened.

SIMPLE EXPLANATION:

Many scientists believe the Moon was formed when an object about the size of Mars slammed into Earth soon after Earth had formed. The tremendous impact flung debris from Earth and the object into space. The debris was trapped in Earth's orbit. Over time, this debris—thousands of pieces of rocks and dust—gradually came together to form one solid object, our Moon.

SOME MORE DETAILS:

Scientists believe this scenario is accurate because of the composition of the Moon. It is like Earth in many ways, but there are a few differences. If the Moon had been part of Earth and then somehow ripped away, it should be made of exactly the same things found on Earth, but it's not. If the Moon had formed someplace else and then was trapped by Earth's gravity, its composition should be much different from Earth's, but it's not.

The impact theory seems to solve both of these problems. The differences in the Moon's composition can be blamed on the debris from the large body that hit Earth. The similarities can be explained by the debris that came from Earth itself.

45. HOW OLD ARE ROCKS ON THE MOON'S SURFACE?

THE ANSWER:

Three to four and a half billion years old.

SIMPLE EXPLANATION:

Apollo astronauts brought back many samples of rock from various parts of the Moon. The oldest samples were found in the lighter-colored, heavily cratered regions. One sample was 4,420,000,000 (4.42 billion) years old.

The youngest rocks were found in the smooth, dark areas called maria (mah'-ree-ah). The youngest sample found was 3,100,000,000 (3.1 billion) years old. This means the Moon has changed its appearance very little in the last 3 billion years.

46. WHY DOES THE MOON CHANGE SHAPE?

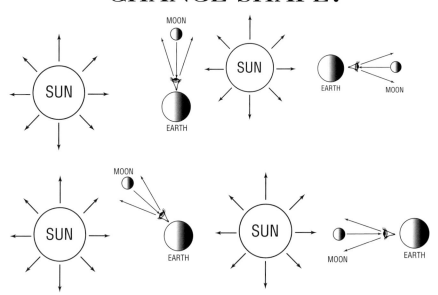

THE ANSWER:

It doesn't, but it seems to change because we see the Sun shining on different parts of the Moon as it travels around Earth.

SIMPLE EXPLANATION:

Some people believe different shapes the Moon goes through, or phases, are caused by Earth's shadow falling on different parts of the Moon at different times. This is not true.

The Moon's changing shape has nothing to do with Earth's shadow. The reason we see the Moon go through phases is that we are seeing different parts of the sunlit side of the Moon as it travels around Earth. (See drawings under Question 52: "Is there a dark side of the Moon?," p. 59.)

SOME MORE DETAILS:

During a Full Moon, it looks as though you are seeing all the Moon. In fact, you are seeing all of the sunlit side. You can see a Full Moon only when it has reached a point in its orbit where Earth lies between it and the Sun.

You don't always see a Full Moon, though. Sometimes you only see half the Moon. What you are really seeing is half of the sunlit side of the Moon and half of the side the Sun is not shining on.

Sometimes the Moon appears as a crescent. It has moved so you can only see a little bit of its sunlit side.

There are times when the sunlit side is facing away from Earth and you can't see the Moon at all. This is called New Moon. At this point, the Moon lies between the Sun and Earth.

After New Moon, as the Moon continues to travel around Earth, the phases start over and we begin to see more of its sunlit side.

Understanding the phases of the Moon is not as easy as it may seem. Just keep in mind that the phases are not caused by Earth's shadow. They are caused by the Moon traveling around Earth, showing us different parts of its sunlit side.

47. WHAT IS A LUNAR ECLIPSE?

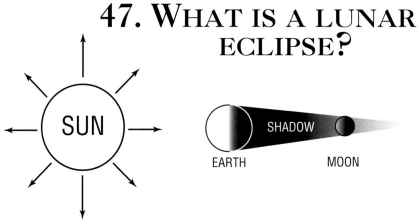

THE ANSWER: A lunar eclipse occurs when Earth's shadow really does fall on the Moon.

SIMPLE EXPLANATION:

During a Full Moon, when Earth is between the Sun and Moon, sometimes the Moon passes through Earth's shadow. If only a part of the Moon passes through Earth's shadow, it creates a partial lunar eclipse. If all of the Moon passes through Earth's shadow, it is a total lunar eclipse.

SOME MORE DETAILS:

Lunar eclipses occur only during a Full Moon. But they do not occur during every Full Moon. Most of the time, the Moon's orbit takes it above or below Earth's shadow.

48. CAN YOU ALWAYS SEE THE MOON?

THE ANSWER:

No. Sometimes you can see the Moon during the evening, and

sometimes you can see it during the day. There is even a period every month when you can't see it at all.

SIMPLE EXPLANATION:

During the New Moon phase, no part of the sunlit side of the Moon can be seen from Earth.

49. WHAT IS THE MAN IN THE MOON?

THE ANSWER:

An imaginary figure made out of the light and dark areas on the Full Moon.

SIMPLE EXPLANATION:

Besides the man in the Moon, stories from all over the world tell of different creatures found on the Moon. Some speak of a dragon, others call it a rabbit, and one story tells of a man who sells cabbages!

Without a telescope, the smooth dark areas, or "maria," and the lighter-colored, heavily cratered regions on the Moon blur together. With your imagination, maybe you can see a man in the Moon

50. DOES THE SAME SIDE OF THE MOON ALWAYS FACE EARTH?

THE ANSWER:

Yes.

SIMPLE EXPLANATION:

Whenever you see the Moon in the sky, you always see the same craters and smooth dark areas. That is because the same side of the Moon always faces Earth.

SOME MORE DETAILS:

Scientists believe that a long time ago the Moon spun around on its axis, so that all sides of the Moon could be seen from Earth. But the Moon is slightly heavier on one side. As the Moon spun around, Earth's gravity tugged on the heavier side, gradually slowing the Moon's spin. Finally, Earth's gravity won, and the heavier side of the Moon now faces Earth all the time.

51. WHAT IS THE "FAR SIDE OF THE MOON?"

THE ANSWER:

The side of the Moon that never faces Earth.

SIMPLE EXPLANATION:

The far side of the Moon cannot be seen from Earth. The only pictures we have of the far side were taken by Apollo astronauts and various satellites that have traveled around the Moon.

SOME MORE DETAILS:

The far side looks quite different from the side of the Moon we can see. There are no smooth dark areas, only a land covered with craters.

52. IS THERE A DARK SIDE OF THE MOON?

THE ANSWER:
Yes and no.

SIMPLE EXPLANATION:
At any given moment half the Moon is light and half is dark. But as the Moon moves around Earth, different portions of its surface go from day to night and vice versa. No portion of the Moon is dark all the time.

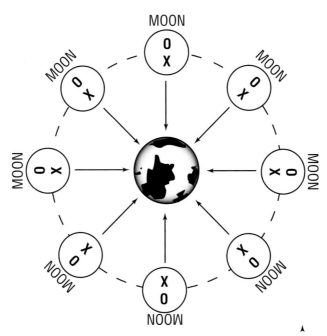

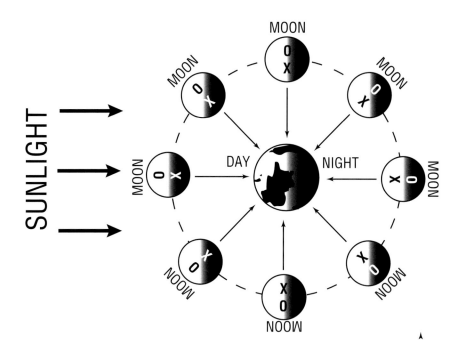

SOME MORE DETAILS:

From the previous questions, we know that the same side of the Moon always faces Earth. In the drawings above, you can see the Moon in eight different parts of its orbit around Earth.

In the first drawing, an "x" marks a crater that can be seen from Earth. An "o" marks a crater on the far side. Notice that as the Moon travels around Earth, the "x" always faces Earth and the "o" always faces away from Earth. The next drawing adds sunlight and shadows to the picture. We cannot see the part of the Moon the Sun is not shining on. If the "x" is in the sunlight, we can see the crater. If the "x" is in the dark, we can't see that crater. As the Moon travels around its orbit, sometimes the "x" is in the sunshine, and sometimes the Sun is shining on the "o." As you can see, every side of the Moon receives sunlight, just at different times.

QUESTIONS ABOUT SPACECRAFT AND SPACE TRAVEL

53. CAN THE SPACE SHUTTLE TRAVEL TO THE MOON?

THE ANSWER:
No.

SIMPLE EXPLANATION:
The space shuttle was not designed to go to the Moon. The shuttle's main engines and the solid rocket boosters are not powerful enough to launch the shuttle to the Moon, nor can the shuttle carry all of the fuel and supplies that would be needed for the long voyage.

SOME MORE DETAILS:
The space shuttle was designed to travel in what is known as low Earth orbit. This means that the highest the space shuttle can travel is about 600 miles (960 kilometers) above Earth's surface. The Moon is, on average, over 230,000 miles (370,000 kilometers) away.

54. DOES THE SPACE SHUTTLE HAVE TO DODGE STARS?

THE ANSWER:
No, just space debris from old satellites.

SIMPLE EXPLANATION:
As stated previously, the space shuttle was designed to travel in low Earth orbit. The highest orbit it is capable of reaching is only about 600 miles (960 kilometers) above the surface of Earth.

The closest star to Earth is the Sun, which lies 93,000,000 miles (149 million kilometers) away from us. The next-closest star, Alpha Centauri, is 25,000,000,000,000 miles (40 trillion kilometers) from Earth. All other stars are much more distant than that.

If you compare the 25,000,000,000,000 miles (40 trillion kilometers) to Alpha Centauri to the 600 miles (960 kilometers) the space shuttle can travel, it becomes apparent that the shuttle doesn't have to worry about running into stars!

SOME MORE DETAILS:

Something the shuttle does have to worry about is running into debris from satellites. If the shuttle ran into something as small as a grain of sand, there could be serious damage to the spacecraft.

Everything in Earth orbit is traveling at extremely high speeds. If a piece of debris hit a window or a less shielded area of the shuttle, it could puncture the spacecraft, depressurize the cabin, and kill everyone on board.

It could do other damage as well. If the debris ripped off some of the protective shielding on the underside of the spacecraft, the shuttle could burn up on re-entry. So you see, space debris is a serious concern. Dodging stars, however, is nothing to lose sleep over.

55. WHY SHOULD WE BUILD A SPACE STATION?

THE ANSWER:

If humans are going to explore outer space, we need to know a lot more about it.

SIMPLE EXPLANATION:

Space, with its micro-gravity and no air to breathe is very different from what we are used to on Earth. The U.S. Skylab missions and the Soviet Soyuz and Mir missions have greatly improved our knowledge, but there is much more we need to understand about the unique environment of space.

Also, while trying to understand the environment of outer space, we gain a better understanding of the environment we live in on Earth.

SOME MORE DETAILS:

The space shuttle allows astronauts to conduct many experiments in the weightlessness of space. Because the shuttle missions are limited to just one or two weeks, the experiments must be kept short. A space station would allow a much longer time for experiments.

The long-term effects of weightlessness on the human body can be studied. The reactions and behavior of crew members in the cramped quarters of the space station over a long period can be studied as well. Information from a space station would be useful in other missions, whether we are developing a colony under the sea, planning a colony on the Moon, or traveling to another planet.

56. IF WE WANTED TO, COULD WE TRAVEL TO THE MOON TOMORROW?

THE ANSWER:

No.

SIMPLE EXPLANATION:

We have been to the Moon, but the last mission took place in 1972, over twenty years ago. The only U.S. rocket powerful enough to take humans to the Moon was the Saturn V. Today, there are only two or three of those rockets around and they are lying on their sides, rusting in front of space museums. If Congress suddenly passed a budget that would fund a mission to the Moon, it would take at least a decade to get there. Why? Because scientists and engineers would have to start nearly from scratch.

We have the technology to go to the Moon. We've been there before. But that technology is nearly thirty years old. Would you want to use a computer that was thirty years old? It would take time to redesign everything to take advantage of today's technology.

SOME MORE DETAILS:

There are some updated plans to go to the Moon. The only thing that is needed is money. But even with money to pay for everything, it would take at least a decade before everything was ready for launch.

57. HOW MANY ASTRONAUTS HAVE WALKED ON THE MOON?

THE ANSWER:

Twelve.

SIMPLE EXPLANATION:

For each of the Apollo missions, there were three astronauts. Two landed on the lunar surface while one remained in the command module. Apollo missions 11, 12, 14, 15, 16, and 17 landed on the Moon. An explosion on board Apollo 13 on its way to the Moon made a lunar landing impossible. Apollo 13 astronauts were able to orbit the Moon only once before they had to return to Earth.

SOME MORE DETAILS:

The astronauts who walked on the Moon were: Neil Armstrong, Buzz Aldrin, Pete Conrad, Al Bean, Alan Shepard, Edgar Mitchell, Dave Scott, Jim Irwin, John Young, Charlie Duke, Eugene Cernan, and Jack Schmitt.

58. CAN WE SEE FOOTPRINTS ON THE MOON?

THE ANSWER:

No, not even through a telescope.

SIMPLE EXPLANATION:

Using the largest telescopes on Earth, the smallest object we can see on the Moon is about a half-mile (1 kilometer) across. The Apollo astronauts wore big boots, but not that big!

59. IF WE BUILD COLONIES ON THE MOON, COULD WE SEE THEM FROM EARTH?

THE ANSWER:

Perhaps, but only with a telescope. It depends on how large the colony is and if it is constructed of material that reflect sunlight.

SIMPLE EXPLANATION:

It is unlikely that we would be able to see the colonies on the Moon.

First of all, the first colonies would be very small. Remember, everything has to be brought from Earth.

Most of the colonies would be underground to protect the inhabitants from meteor impacts and radiation from the Sun. What little is left above the surface would probably be mining equipment or extra supplies. None of these things would be big enough to see from Earth. Maybe in the distant future, cities on the Moon will be large enough to see from Earth without a telescope. They will have to be very large, though—at least 50 miles (80 kilometers) across.

SOME MORE DETAILS:

Because the Moon is so far away from us, we can see very little detail on its surface with the naked eye—only features greater than 50 miles (80 kilometers) across. You can't see anything smaller than 50 miles (80 kilometers) unless you use a telescope.

60. WHY HAVEN'T HUMANS TRAVELED TO OTHER PLANETS?

THE ANSWER:

Because it would take a really long time to get there.

SIMPLE EXPLANATION:

It would take a long time, a lot of fuel, food, and air to breathe, and an extremely large spacecraft to carry everything. But by far the most important thing that would be required is money. Such a trip would be far too expensive for any nation to undertake right now or in the foreseeable future.

SOME MORE DETAILS:

The major problem with interplanetary travel, besides the money, is the vast distance involved. It took the Apollo astronauts three solid days of travel to get to the Moon. That's pretty far, especially if you think of it in terms of traveling in a car. Could you travel in a car for three days without stopping for food or gasoline or to stretch your legs?

It takes three days to get to the Moon and three days to get back. If you want to spend time on its surface, it takes even longer. Humans have done that. If you wanted to go to Venus or Mars, it would take four to six months to get there, and even longer to get back. And those are the two closest planets. It would take years to reach Jupiter, Saturn, and the other outer planets.

61. SHOULD WE TRAVEL TO MARS?

THE ANSWER:

Yes.

SIMPLE EXPLANATION:

It is human nature to explore, and of all the other planets in our solar system, Mars has the greatest potential to support human life.

SOME MORE DETAILS:

While no planet in the solar system other than Earth could support life as we know it, Mars comes the closest.

Even though astronauts would have to deal with below-freezing temperatures, no air to breathe, and almost no atmosphere, scientists already have plans for traveling to the Red Planet.

Several ideas have been suggested on how a Mars mission could be accomplished. One idea is to send in advance the food and supplies needed for the stay on the planet and the return trip. Not everything will have to be crammed aboard the crew ship, and the supplies will already be there when the astronauts arrive.

Once on the surface, scientists think astronauts will be able to use the Martian soil for some of their building supplies. From the information sent back from Viking 1 and 2, scientists think that astronauts will be able to make plaster of paris, ceramics, a type of cement, glass, and maybe even blasting explosives out of the Martian soil. This will cut down on the amount of material that will have to be brought from Earth.

The key factor in deciding whether we travel to Mars is money.

Scientifically and technologically, we are capable of traveling to Mars. Unfortunately, everything costs money—and a mission this size costs a lot of money.

62. CAN WE TRAVEL TO THE STARS?

THE ANSWER:

No.

SIMPLE EXPLANATION:

The stars you see at night are relatively distant. Alpha Centauri is the closest of those stars. It is 25,000,000,000,000 miles (40 trillion kilometers) away from us. If you got in your car and magically drove to Alpha Centauri, going 65 miles per hour, it would take you 44 million years to get there.

Voyager 2, the fastest spacecraft ever launched, is flying out of our solar system at a speed of about 40,000 miles per hour. If you could travel as fast as Voyager 2, it would still take you over 70,000 years to get to Alpha Centauri. Remember, that is just to the closest star. For now, we are stuck in our own solar system.

63. DO ANY OF THE SPACECRAFT WE SEND TO THE PLANETS RETURN TO EARTH?

THE ANSWER:

Almost never.

SIMPLE EXPLANATION:

Almost all of the spacecraft that are launched to other worlds will never return to Earth. The few that do return to Earth are special "sample return" missions. These special missions have returned samples from the Moon and will one day return soil samples from Mars.

SOME MORE DETAILS:

The reason more spacecraft do not return to Earth is gravity. To escape Earth's gravitational field, every spacecraft we send to the planets has to use powerful rockets. The bigger the object you want to launch, the more powerful your rockets have to be. Every pound of weight added to a spacecraft requires that much more lift from a rocket. So the heavier the spacecraft, the larger the rocket.

Scientists try to put as many instruments on a spacecraft as they can without increasing the weight too much. Instruments that would bring the spacecraft back to Earth would weigh a lot and would require a larger rocket to launch.

64. HOW DO WE GET ALL THOSE PRETTY PICTURES OF THE PLANETS?

THE ANSWER:

The same way you receive music through your radio or a picture on your television screen—by using radio waves.

SIMPLE EXPLANATION:

Scientists use these radio waves to send messages to and receive messages from a computer on a spacecraft. All spacecraft have a computer. (Remember, these are robot spaceships with no humans on board.) Scientists control cameras on the spacecraft by talking to the onboard computer using radio signals.

SOME MORE DETAILS:

Using radio signals, scientists tell the computer to aim a camera at a certain point. The same radio signals tell the computer how long the exposure is supposed to be and where to aim the camera to take the next picture.

Once an image is taken, there is no one on the spacecraft to send the film out for processing! There isn't even "film" on board. The computer remembers the image as a series of numbers and sends those numbers back to Earth as a radio signal. A computer on Earth receives the radio signal and converts the numbers back into an image. That's how we get all the pictures of the planets.

QUESTIONS

ABOUT

STARS

65. Is it true that stars twinkle and planets don't?

The Answer:

No and yes.

Simple Explanation:

The saying that stars twinkle and planets don't is not correct. Neither stars nor planets actually "twinkle." However, we notice apparent changes in stars more than we do in planets. So, the second answer is yes, most of the time stars seem to twinkle and planets don't.

Some More Details:

The twinkling effect, which causes a star to shimmer or sparkle, is not caused by the star or planet. It is caused by good ol' Mother Earth. Or more specifically, Earth's atmosphere. Light from a star or planet doesn't begin to twinkle until it passes through the turbulent air that we breathe.

Stars, with the exception of our Sun, are extremely distant. Only a small amount of their light reaches us. This light travels into our eyes and activates one of the light-detecting "rods." When a rod is activated, we can see the star. Sometimes, before the starlight reaches our eyes, its path is disturbed by Earth's turbulent atmosphere. This causes the light to bounce from one rod to another, turning off one and turning on another. Our brain picks this movement up and the stars appear to twinkle.

A planet, on the other hand, is much closer to us and much brighter than the distant stars. It forms a much larger image on our retina. The image of a planet is large enough to turn on several light-detecting rods at once, instead of just one. When several rods are turned on at once, if one or two are turned off, there is not a noticeable change. In other words, the image of a planet is large enough that you don't notice the slight disturbance caused by the atmosphere.

On some nights, however, the atmosphere is so turbulent that everything in the sky twinkles. At those times, it's best to put away your telescope and read a book.

66. WHAT IS A CONSTELLATION?

THE ANSWER:

A collection of stars that we have formed into an imaginary pattern like Leo the Lion or Orion the Hunter.

SIMPLE EXPLANATION:

A long time ago, people played connect-the-dots with the stars and formed patterns in the sky. These patterns are the constellations we are familiar with today.

SOME MORE DETAILS:

Most people are familiar with the Big Dipper, a group of stars that is part of the constellation Ursa Major (the Great Bear).

The seven stars of the Big Dipper are connected only by our imagination. These stars actually vary greatly in their brightness and distance from Earth. From Earth, they appear to form a dipper. From a different position—say from a planet orbiting another star—the seven stars would form a different shape. We would see them from a different perspective.

67. HOW MANY CONSTELLATIONS ARE THERE?

THE ANSWER:

Eighty-eight

SIMPLE EXPLANATION:

There are a total of 88 constellations in the sky. This includes all of the constellations in the Northern and Southern Hemispheres.

SOME MORE DETAILS:

A person living in the Northern Hemisphere cannot see all 88 constellations. Most of the southern constellations are below the horizon throughout the year, hidden from view.

68. DO ASTRONOMERS STUDY CONSTELLATIONS?

THE ANSWER:

No.

SIMPLE EXPLANATION:

Constellations are great for finding your way around the sky and for telling stories around a campfire. In the old days, sailors like Columbus and Magellan used constellations to find their way on the seas. But today, astronomers don't use them much.

SOME MORE DETAILS:

Constellations are huge imaginary figures containing many of the bright stars in the sky. Astronomers study individual stars and physical groups of stars such as star clusters and galaxies. They study objects in the sky, not patterns of stars on the sky. Also, astronomers have divided the sky into a coordinate system, so they no longer need the constellations to help them find the fainter stars.

69. CAN YOU SEE A CONSTELLATION THROUGH A TELESCOPE?

THE ANSWER:

No.

SIMPLE EXPLANATION:

When you look through a telescope, you see only a few stars in a small area of the sky. Stars in a constellation are stretched across a large portion of the sky. Even the smallest constellations are far too large to view in a telescope.

70. WHY DO THE STARS ONLY COME OUT AT NIGHT?

THE ANSWER:

Because the light from the Sun drowns them out during the day.

SIMPLE EXPLANATION:

Stars cover the sky at all times, even in the brightest part of the day. However, the Sun doesn't let us see them. Only when our side of Earth turns away from the Sun can we see stars.

SOME MORE DETAILS:

Have you ever had someone shine a flashlight in your eyes? The flashlight is so bright that you can't see who is committing the nasty deed. However, if you turn your back to the flashlight, you can see all that was behind you.

The Sun acts just like the flashlight, blinding us so we can't see the other stars around it. But as Earth turns its back on the Sun, we can see the stars at night.

71. HOW MANY STARS CAN YOU SEE WITH YOUR EYES ALONE?

THE ANSWER:

What may seem like billions of stars on a dark, clear night is only a little over 5,500.

SIMPLE EXPLANATION:

There are approximately 200 billion stars in our Galaxy, the Milky Way. But even on the darkest, clearest night, human eyes can pick out only 5,500 individual stars. And that's if you stay up all night!

Human eyes are not sensitive enough to see the other 199+ billion stars. That is why astronomers use telescopes.

72. WHAT IS THE NORTH STAR?

THE ANSWER:

The star that always can be found in the north. This star's real name is Polaris.

SIMPLE EXPLANATION:

Have you ever seen someone spin a basketball on her finger? As the basketball spins around on her finger, the middle of the basketball spins around in a big circle. The area under her finger is spinning, too, but in very tiny circles.

Now think of Earth as the basketball. Instead of spinning around on a finger, it spins around on its axis. Earth's axis is represented by the North and South Poles. The person's finger represents the South Pole.

The North Pole is pointed in the direction of the star Polaris. As Earth turns, the North Pole turns as well, but in very tiny circles. So while Earth periodically rotates away from all of the other stars, Polaris seems to stay fixed in our skies, always pointing the way north.

73. IS THERE A SOUTH STAR?

THE ANSWER:

No.

SIMPLE EXPLANATION:

No bright star lines up with Earth's South Pole, so there is not a South Star. The fact that there is a star lined up with the North Pole is coincidence.

74. WILL THE NORTH STAR ALWAYS SHOW US THE WAY NORTH?

THE ANSWER:

No.

SIMPLE EXPLANATION:

Polaris, our North Star, will guide us north for the rest of our lifetimes and that of our great-great-grandchildren as well. So for us, Polaris could be considered a permanent North Star. But in several thousand years, that will change.

SOME MORE DETAILS:

Polaris is the North Star because Earth's North Pole points in its direction. But Earth's North Pole wobbles slightly as Earth travels around the Sun. This movement is so slight that it takes 26,000 years to complete a wobble. But that wobble, called precession, slowly moves the North Pole in a huge circle.

Some two thousand years ago the North Pole pointed to the star Thuban, in the constellation Draco. In 12,500 years, the North Pole will point near the star Vega in the constellation Lyra. In 26,000 years the North Pole will once again point to Polaris.

75. IS THE NORTH STAR THE BRIGHTEST STAR IN THE SKY?

THE ANSWER:

No, not by a long shot.

SIMPLE EXPLANATION:

Since just about everyone has heard about the North Star, many people think it must be the brightest, most impressive star in the sky. That's not true.

Polaris is just an average star. What makes it special is its location in the sky. If it were not lined up with Earth's North Pole, Polaris would simply be known as the brightest star in the faint constellation Ursa Minor (the Little Bear).

SOME MORE DETAIL:

The brightest star in the night sky is called Sirius. It is located in the constellation Canis Major (the Big Dog) and is sometimes called the Dog Star. The brightest star in the daytime sky is, of course, our Sun.

76. WHAT IS A LIGHT-YEAR?

THE ANSWER:

A measure of distance.

SIMPLE EXPLANATION:

In astronomy, distances are so great that astronomers invented a unit to describe trillions of miles. If they didn't, they would have to spend most of their time writing zeroes. It would be like measuring a football field in inches.

So astronomers invented a light-year. One light-year is equal to the distance that light travels in a year. You may not realize that light takes time to travel anywhere. Whenever you turn on a light switch, the light seems to be right there. You don't have to wait.

Well, light travels extremely fast. In fact, scientists believe that nothing can travel faster than the speed of light. Light travels 186,000 miles (300,000 kilometers) every second. If you lived 186,000 miles from the nearest power station, it would take one second for the light to reach your house after you hit the switch.

SOME MORE DETAILS:

If light travels 186,000 miles every second, how far does it travel in an entire year?

There are 60 seconds in a minute, 60 minutes in an hour, 24 hours in a day, and 365 days in a year. Multiply all of those together and you get the distance light travels in one year. 186,000 x 60 x 60 x 24 x 365 = 5,900,000,000,000 miles/year. Even traveling that fast, it takes time for light to travel the huge distances in space. Within our solar system, light takes 8.5 minutes to travel from the Sun to Earth. To get to Pluto, it takes light 5.5 hours to travel from the Sun. To travel to the next nearest star, Alpha Centauri, it takes light from our Sun 4.2 years.

77. WHAT IS THE CLOSEST STAR TO EARTH?

THE ANSWER:

The Sun, but most people forget that the Sun is a star.

The next-closest star to Earth is Alpha Centauri, the brightest star in the southern constellation Centaurus. This star is 4.2 light-years or 25,000,000,000,000 miles (40 trillion kilometers) away from us.

SIMPLE EXPLANATION:

Alpha Centauri lies too far south in the sky to be seen from most of the Northern Hemisphere. It may be the closest star in the night sky, but it is not the brightest. Two other stars, Sirius and Canopus, are farther away but are brighter than Alpha Centauri.

SOME MORE DETAILS:

Through a telescope, Alpha Centauri is a triple star. The two larger stars of the system orbit very close to each other. A third, fainter star called Proxima Centauri is actually somewhat closer to us than Alpha and orbits the brighter pair.

78. HOW DOES A STAR FORM?

THE ANSWER:

A star forms from a huge cloud of gas and dust.

SIMPLE EXPLANATION:

These clouds, called nebulae, are made up mostly of hydrogen. The hydrogen gas is not spread out evenly within the cloud. It clumps together in places.

A cloud, or nebula, gradually attracts more hydrogen. As hydrogen collects, the cloud becomes more massive. As the mass increases, so does its gravity, attracting still more gas.

While the cloud grows in size, the gas at its center is squeezed by the strong gravity. As gravity squeezes the hydrogen atoms together, the temperature begins to rise. It's like being in a small room. The more people that crowd into the room, the hotter the room gets.

Finally, the temperature in the center is so hot and the pressure of all of the hydrogen atoms being squeezed together is so great that nuclear fusion begins. Four hydrogen atoms are fused into one helium atom, releasing a tremendous amount of energy.

When nuclear fusion begins, a star is born. Eventually, the pressure from the reactions pushing out from the center of the star equals the gravity pushing in on the star, and it is balanced. The star will continue to emit energy as long as nuclear reactions continue in its core.

79. WILL A STAR LIVE FOREVER?

THE ANSWER:

No.

SIMPLE EXPLANATION:

In some ways, stars are just like humans.

They are born, live out their lives, and die. But instead of living for seventy years, like some humans, they live for millions or billions of years.

After living an extremely long life (by human standards), a star will run out of fuel and die. If the star is small, it uses up its fuel quietly and slowly fades away. If the star is many times larger than our Sun, though, it uses all of its fuel quickly and explodes in a supernova. Nothing lives forever.

80. HOW LONG DO STARS LIVE?

THE ANSWER:

When a star is born, its size determines how long it will live. The smaller the star, the longer it will live.

SIMPLE EXPLANATION:

The small, low-mass stars fuse hydrogen into helium very slowly. These cool, red stars burn for trillions of years before they use all of the hydrogen in their core.

Medium-sized stars, like our Sun, burn faster. Because they are more massive than the red stars, the pressures in their cores are greater, causing nuclear reactions to occur more quickly. Medium-sized stars live for a few billion years.

The larger, more massive stars have the shortest lives. These stars are the hottest and the brightest in our skies, and they die within a few million years. When it comes to stars, the bigger you are, the shorter your life.

81. WHAT IS A RED GIANT?

THE ANSWER:

A big, old, red star.

SIMPLE EXPLANATION:

Stars are not born as red giants. Instead, a star like our Sun becomes a red giant when it has used up most of its fuel and begins to grow in size.

As a star grows in size, its surface begins to cool off and turn red in color. Hence, the "red" part of the term. (In star colors, cool = red, warm = yellow, and hot = blue/white) When our Sun becomes a red giant (a few billion years from now) it will grow until it swallows Mercury and Venus and extends almost all the way out to Earth. Hence the "giant" part of the term.

82. WHAT IS A WHITE DWARF?

THE ANSWER:

A "dead" star about the size of Earth.

SIMPLE EXPLANATION:

After a star like our Sun uses up all of its fuel, it shrinks into a small sphere, about the size of Earth. After it shrinks, it becomes a white dwarf.

A white dwarf remains very hot and bright for a long time as it cools off. But no new nuclear reactions are going on in its core. So a white dwarf is considered a dead star. Eventually it will cool even more and fade to black.

SOME MORE DETAILS:

Matter that was once spread out inside an entire star is now compressed into a sphere the size of a small planet. This makes the star so dense that a teaspoon of material from a white dwarf would weigh a couple of tons.

83. WHAT IS A SUPERNOVA?

THE ANSWER:

A huge explosion that marks the death of a star many times larger than our Sun.

SIMPLE EXPLANATION:

When a supermassive star—one that is over three times larger than the Sun—uses up its fuel, the star begins to collapse. The collapse happens so fast that the star's material collides with itself in the center. A tremendous rebound occurs and the star tears apart.

A supernova releases so much energy that the star's brightness increases one hundred millionfold. A star ordinarily not visible from Earth can suddenly become the brightest star in our sky and can even be seen during the day.

84. WHAT IS A PULSAR?

THE ANSWER:

A very small, dead star spinning incredibly fast. As it spins, it seems to be sending a series of radio pulses to Earth. The word "pulsar" is short for pulsating star.

SIMPLE EXPLANATION:

When astronomers discovered the first pulsar in 1967, they had no idea what it was. Using a radio telescope, they found that if they aimed the telescope at a specific spot in the sky, they recorded a series of "blips." These radio blips were so regular that the astronomers didn't think they could be produced by anything natural. They called the unknown object LGM, short for Little Green Men.

It turns out that pulsars don't have anything to do with little green men. Instead, they are formed after the death of massive stars. A pulsar is the rapidly spinning remnant of a dead star, located within the debris of a supernova.

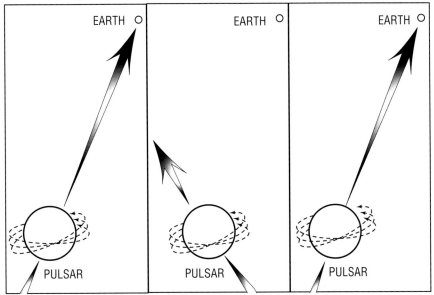

SOME MORE DETAILS:

The pulses from a pulsar are caused by its fast rotation and the fact that its magnetic north or South Pole is not pointed directly at Earth.

Before we tackle the problem of the pulsar, let's examine something more familiar. Imagine a friend has a flashlight pointed toward you. Now your friend begins to spin around. You see a flash of light every time your friend turns the flashlight in your direction. When your friend is facing away from you, you don't see anything.

Now imagine your friend is a pulsar and the flashlight he's holding is the pulsar's magnetic North Pole. Pulsars have strong magnetic fields, tremendously stronger than Earth's. Charged particles trapped in this field release radiation that escapes the pulsar at its magnetic north and magnetic South Poles. We detect this radiation as a blip every time the north or south magnetic pole points in our direction.

There is something else you can learn about pulsars. Let's go back to your friend with a flashlight. You can tell how fast your friend is spinning by counting how long it is between flashes of light. The faster he spins, the faster you will see the blips of light. Your friend may be able to spin around once every one or two seconds, but he will probably get dizzy and have to stop. Pulsars have been found to rotate on their axes anywhere from once every 1 second to 30 times a second.

85. WHAT IS A NEUTRON STAR?

THE ANSWER:

A neutron star is the dense remnant of a dead star. A pulsar is a rotating neutron star.

SIMPLE EXPLANATION:

A neutron star forms after a supernova explosion. It is dead because no nuclear reactions are occurring in its core. But wait a minute! A supernova occurs when a huge star rips itself apart. How can anything be formed after such a tremendous explosion?

Sometimes when a supernova explosion takes place, the core of the star survives. With the rest of the star destroyed, the core begins to collapse in on itself. The leftover material collapses into a sphere about ten miles across. This is a neutron star.

It is also very dense. A teaspoon of material from a neutron star would weigh several million tons.

SOME MORE DETAILS:

Think about this. All the material that was in a star many times larger than our Sun, is now crammed together inside a ball smaller than the state of Rhode Island. Now, that's dense!

86. WHAT IS A BLACK HOLE?

THE ANSWER:

A place where gravity is so strong that nothing—not even light—can escape.

SIMPLE EXPLANATION:

Gravity is what keeps us attached to Earth.

If you throw a ball up in the air, gravity will pull the ball down until it hits the ground. If you could throw the ball fast enough, though, it would escape Earth's gravity. You would have to throw the ball really hard, faster than 25,100 miles per hour (40,320 kilometers per hour). Even the best pitchers in baseball can only throw about 100 miles (160 kilometers) per hour. Apollo astronauts had to travel faster than 25,100 miles per hour (40,320 kilometers per hour) before they could leave Earth and travel to the Moon.

The more massive the object, the stronger its gravity. Jupiter is more than eleven times the diameter of Earth and has a much stronger gravity. To escape from Jupiter, you would have to travel faster than 133,000 miles per hour (214,560 kilometers per hour).

A black hole is so massive that its gravity is the strongest we know of. The gravity of a black hole pulls on things so hard that you would have to travel faster than 186,000 miles (300,000 kilometers) per second to escape. 186,000 miles per second is the speed of light. Nothing, however, can travel faster than light.

Because nothing can travel faster than the speed of light, nothing can escape a black hole. Everything that is swallowed up by a black hole, including light, will remain inside it.

87. WHAT IS A SINGULARITY?

THE ANSWER:

The actual black hole, the point where everything ends up.

SIMPLE EXPLANATION:

A singularity could be called the center of a black hole. This is where the laws of physics break down. A singularity is the point where there is incredible mass that takes up no volume. Everything that enters a black hole will end up here.

88. WHAT IS AN EVENT HORIZON?

THE ANSWER:

An event horizon surrounds a singularity and could be considered the "surface" of the black hole.

SIMPLE EXPLANATION:

An event horizon marks the boundary of a black hole. Once something crosses this boundary, it must be able to travel faster than the speed of light to escape. Since nothing can travel that fast, everything that crosses the event horizon is trapped in the black hole.

Once inside the event horizon, everything will end up within the singularity. An object is not trapped by the black hole until it crosses the event horizon. The more mass contained in the black hole, the larger the event horizon.

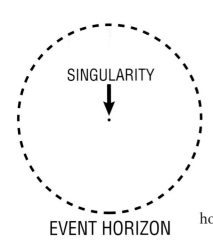

BLACK HOLE

SINGULARITY

EVENT HORIZON

89. How do black holes form?

The Answer:

Sometimes with the death of a supermassive star, and sometimes in the center of galaxies. No matter how the process is started, the result is the same. Gravity takes over and collapses everything within reach into one tiny point.

Simple Explanation:

After a supernova explosion, the gravity of the stellar debris can cause it to collapse in on itself. If there is enough material, the debris will collapse past the point of a neutron star. In fact, the gravity continues to squeeze the debris together, forcing the collapse, until everything that was left over from a star has been squeezed into a single point. This point (a singularity) and a surrounding border (an event horizon) make up the black hole.

The size of the black hole depends on how much mass is inside. The greater the mass, the larger the black hole. Large, supermassive black holes are found in the center of galaxies. Instead of containing all the matter from one star, these black holes contain the matter from many stars. Sometime during the formation of a galaxy, stars crowded within the central core collapse in upon themselves, forming a huge black hole.

You may wonder how we can talk about various sizes of black holes if the matter entering a black hole ends up inside a singularity that doesn't take up any space. When astronomers discuss the size of a black hole, they are referring to the size of the event horizon, or surface. The event horizon marks the border where the gravity is strong enough to overcome the speed of light. As more mass is added to the black hole, its gravity increases. The stronger the gravity, the farther away its effects are felt.

SOME MORE DETAILS:

If this sounds confusing, don't worry. Even the laws of physics break down at this point. Physics can explain all of the stages involved in forming a black hole. But once a black hole is formed, physics as we know it can't explain the existence of an object of incredible mass that takes up no space.

90. SINCE BLACK HOLES ARE BLACK, HOW DO YOU FIND THEM?

THE ANSWER:

Astronomers can't see black holes, but using special telescopes they can detect x-rays that may be coming from material that is near a black hole but hasn't yet been drawn inside it. With the Hubble telescope, astronomers have actually seen some of this material.

SIMPLE EXPLANATION:

Since nothing can escape a black hole, there is no way we can find one directly. We can see stars because of the light they emit. But black holes trap light. They trap everything inside the event horizon, which makes them impossible to see.

But they can't swallow everything at once. Once material is caught up by the gravity of the black hole, it has to wait in line to be swallowed. It orbits the black hole, spiraling in toward the event horizon faster and faster, like water being pulled down a drain as you empty a bathtub.

This material surrounding a black hole forms a flattish shape called an accretion disk. As this material spirals into the black hole, it heats up to tremendous temperatures. In fact, it moves so fast and is so hot that it emits x-rays. Since the x-rays are emitted outside the event horizon, they can escape. Astronomers can detect these x-rays. If they detect a strong x-ray source and can't see a visible object emitting those x-rays, they may be "seeing" an accretion disk around a black hole. Astronomers have observed several x-ray sources that could indicate possible black holes.

Telescopes on Earth have not been able to see any accretion disks because of the difficulty in detecting these small, elusive objects. The x-rays coming from the disk are much easier to spot than any visual light that may be released. However, in May 1994 the Hubble Space Telescope, floating high above Earth's murky atmosphere, was able to visually detect an accretion disk surrounding a black hole in the spiral galaxy M87. This was the first sighting of its kind.

91. WHAT WOULD HAPPEN IF OUR SUN BECAME A BLACK HOLE?

THE ANSWER:
Our Sun is not massive enough to become a black hole.

SIMPLE EXPLANATION:
Even if our Sun did suddenly collapse into a black hole, the planets would continue to follow their orbits.

SOME MORE DETAILS:
The tricky part here is mass. As long as the Sun's mass does not change as it shrinks into a black hole, everything else in our solar system would continue to orbit as if nothing happened.

The planets are held in their orbits by the gravity of the Sun. The strength of the Sun's gravity is determined by the Sun's mass (the amount of stuff inside the Sun). If the mass stays the same, the gravity will stay the same, even if that mass shrinks in size.

The effects of strong gravity around a black hole occur only within a few hundred miles of the event horizon. A few hundred miles may not sound close, but remember, we are dealing with huge distances. Mercury, the closest planet to the Sun, would be over 36 million miles (58 million kilometers) from the black hole.

92. COULD YOU TRAVEL THROUGH A BLACK HOLE?

THE ANSWER:
Not a chance! You would be toast long before you got close!

SIMPLE EXPLANATION:
If you could magically travel to a black hole—remember, humans can't even travel to the next planet yet!—you would be killed by many different things before you even reached the black hole.

SOME MORE DETAIL:
Within a few hundred miles of a black hole, its strong gravity would rip you apart. The x-rays from the accretion disk would fry you and the heat would melt even the toughest metals. After you had been ripped apart, fried and melted, what remained would be sucked into the black hole forever.

93. CAN WE SEND A SPACESHIP THROUGH A BLACK HOLE?

THE ANSWER:

No.

SIMPLE EXPLANATION:

Once again, anything trying to travel through a black hole would be completely and utterly destroyed before it even got to the black hole. And once in the black hole, there is no way anything can escape.

SOME MORE DETAILS:

The nearest possible black hole to Earth is several thousand light-years away. Right now, it would take humans thousands of years to travel one light-year. There is no way we could even come close to a black hole. We can only study them from afar.

94. WHERE DOES EVERYTHING GO AFTER IT ENTERS A BLACK HOLE?

THE ANSWER:

It stays in the black hole.

SIMPLE EXPLANATION:

Everything that enters a black hole remains inside the black hole.

Even though everything disappears from our view as it enters the event horizon, astronomers know that everything must still be inside the black hole.

How can they be sure the stuff is still there? If it weren't there, there wouldn't be a black hole. If everything that was swallowed by a black hole passed through into another dimension, another time, or another part of space, pretty soon the black hole wouldn't have any mass left. As it lost mass, the gravity of the black hole would weaken. Without a strong gravity, things would begin to escape and it would no longer be a black hole.

SOME MORE DETAILS:

The fact that the laws of physics break down when describing a black hole makes them a fascinating topic. There are many things in the universe that we don't understand completely. Black holes are just one of them. Scientists know enough about black holes to realize we may never completely understand these celestial garbage disposals.

95. WILL BLACK HOLES DEVOUR THE UNIVERSE?

THE ANSWER:

No.

SIMPLE EXPLANATION:

First of all, black holes are rare. They only form within the centers of large galaxies or after the death of a supermassive star. For every one galactic black hole there are billions of stars. Of those billions of stars, only a very few have enough mass to go supernova. And not all of those stars will form black holes when they explode.

Second, the extremely strong gravity of a black hole affects only things close to it, (within a few hundred miles after a supernova explosion, a few million miles in the galactic center). Outer space is very empty. In astronomical terms, a few hundred miles is nothing at all. Mercury is considered to be close to the Sun and it lies at a distance of 36 million miles (58 million kilometers). The next nearest star to us is over 25 trillion miles (40 trillion kilometers).

So the influence of a black hole can only be felt if you are almost on top of it. That, along with the fact that black holes are very rare, makes the universe a very safe place for those of you worried about being swallowed by a black hole.

QUESTIONS ABOUT GALAXIES

96. WHAT IS A GALAXY?

THE ANSWER:

A huge grouping of stars, planets, star clusters, nebulae, black holes and everything else astronomical.

SIMPLE EXPLANATION:

Just as a house contains a living room, dining room, bedroom, bathroom, etc., a galaxy contains stars, planets, moons, nebulae, black holes, and much more. Although it contains many individual items, a galaxy is considered to be the largest astronomical body. Galaxies themselves vary in size, yet even the smallest is much larger than anything discussed so far in this book.

97. WHAT IS THE MILKY WAY?

THE ANSWER:

Our home Galaxy.

SIMPLE EXPLANATION:

Our solar system is just a small part of the Milky Way Galaxy. The Milky Way is a spiral galaxy that contains 200 billion stars.

A spiral galaxy is a flat circular disk with a huge sphere of stars in the center. When you look at a spiral galaxy face on, it looks like a pinwheel. You can see spiral arms within the disk. When you see it edge on, the disk is very thin and the central ball appears very bright. Our solar system lies in the disk of the Galaxy, on the outer fringes of one of the spiral arms.

SOME MORE DETAILS:

Remember that one light-year is 5.9 trillion miles (9.4 trillion kilometers). Our Galaxy is 100,000 light-years in diameter and about 3,000 light-years thick. The farthest we have sent a spacecraft is just beyond the orbit of Pluto, slightly over 4 billion miles (6.4 billion kilometers) away.

98. WHEN CAN I SEE THE MILKY WAY?

THE ANSWER:

Any clear night.

SIMPLE EXPLANATION:

Our Milky Way is not just the hazy streak of light you can see in the late summer and early autumn night sky. If you don't use a telescope or binoculars, all but three of the objects you see in the night sky are located in our Milky Way Galaxy. Every star, star cluster, planet, and moon is part of the Milky Way.

The hazy streak in the late summer night sky is made up of the more distant stars of our Galaxy. These countless stars are too distant to be seen as pinpoints of light. They all blur together to form the hazy cloud. With a pair of small binoculars, you can begin to make out some of the thousands of individual stars.

SOME MORE DETAILS:

In the Northern Hemisphere there is one object you can see without a telescope that does not belong in our Galaxy. If you know where to look, on a dark, clear night you can see the Andromeda Galaxy. This faint, fuzzy blob is an entirely separate galaxy that contains its own stars, star clusters, nebulae, black holes, and much more. The

Andromeda Galaxy is 2.2 million light-years away from us and is the most distant object you can see with the naked eye.

In the Southern Hemisphere you can see two galaxies, the Large and Small Magellanic Clouds, without a telescope. These two galaxies are companions to our own Milky Way.

99. DO WE HAVE A PICTURE OF OUR GALAXY?

THE ANSWER:

We don't have a picture showing the entire Galaxy from a distance.

SIMPLE EXPLANATION:

We live inside the Milky Way Galaxy. To take a picture of the whole Galaxy, we would have to travel thousands of light-years away from it, just as you would have to go far enough away from your house to get the whole building in a photograph. Right now, human beings can't even travel a few million miles to the next planet.

100. ARE ALL GALAXIES SHAPED LIKE A PINWHEEL?

THE ANSWER:

No. Though pinwheel, or spiral, galaxies may be the prettiest to look at, they are not the only galaxies in the universe. There are elliptical galaxies, irregular galaxies, and peculiar galaxies as well.

SIMPLE EXPLANATION:

The largest and smallest known galaxies are both elliptical.

Elliptical galaxies are huge balls of stars with no spiral arms. They range from an almost perfect sphere to an extreme oval shape.

Irregular galaxies don't have any particular pattern. The Large and Small Magellanic Clouds, the Milky Way's two companions, are both irregular galaxies.

Peculiar galaxies are just that—peculiar. Something strange is happening in these galaxies, and scientists aren't sure what. A peculiar galaxy may have a spiral or an elliptical shape, but it has huge amounts of energy shooting out of it.

101. HOW MANY GALAXIES ARE THERE?

THE ANSWER:
There are over 100 billion other galaxies.

SIMPLE EXPLANATION:
There are more galaxies than there are stars in our sky. From Earth, we can only see three galaxies without a telescope; the Andromeda Galaxy and the Large and Small Magellanic Clouds. The others are too faint for the human eye to see.

Astronomers have photographed thousands of distant galaxies.

Each one of these contains billions of stars. In other words, our universe is a vast place that we will never be able to completely explore.